MW00380796

Cram101 Textbook Outlines to accompany:

Probability and Statistics

DeGroot and Schervish, 3rd Edition

An Academic Internet Publishers (AIPI) publication (c) 2007.

You have a discounted membership at www.Cram101.com with this book.

Get all of the practice tests for the chapters of this textbook, and access in-depth reference material for writing essays and papers. Here is an example from a Cram101 Biology text:

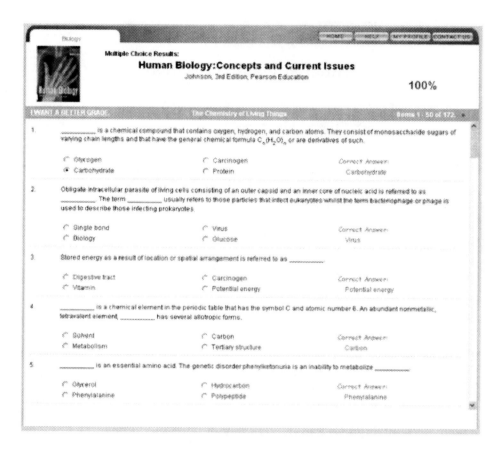

When you need problem solving help with math, stats, and other disciplines, www.Cram101.com will walk through the formulas and solutions step by step.

With Cram101.com online, you also have access to extensive reference material.

You will nail those essays and papers. Here is an example from a Cram101 Biology text:

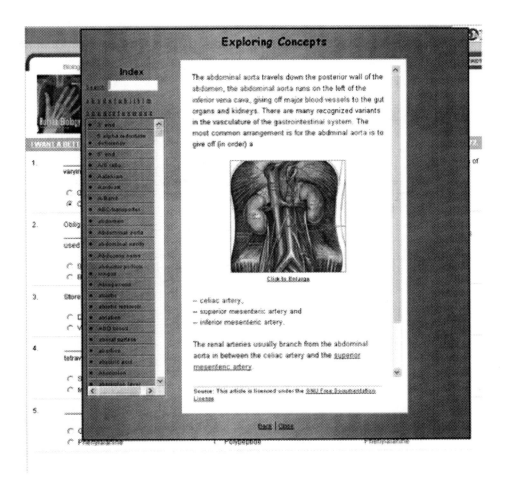

Visit **www.Cram101.com**, click Sign Up at the top of the screen, and enter DK73DW1215 in the promo code box on the registration screen. Access to www.Cram101.com is normally $9.95, but because you have purchased this book, your access fee is only $4.95. Sign up and stop highlighting textbooks forever.

Learning System

Cram101 Textbook Outlines is a learning system. The notes in this book are the highlights of your textbook, you will never have to highlight a book again.

How to use this book. Take this book to class, it is your notebook for the lecture. The notes and highlights on the left hand side of the pages follow the outline and order of the textbook. All you have to do is follow along while your intructor presents the lecture. Circle the items emphasized in class and add other important information on the right side. With Cram101 Textbook Outlines you'll spend less time writing and more time listening. Learning becomes more efficient.

Cram101.com Online

Increase your studying efficiency by using Cram101.com's practice tests and online reference material. It is the perfect complement to Cram101 Textbook Outlines. Use self-teaching matching tests or simulate in-class testing with comprehensive multiple choice tests, or simply use Cram's true and false tests for quick review. Cram101.com even allows you to enter your in-class notes for an integrated studying format combining the textbook notes with your class notes.

Visit **www.Cram101.com**, click Sign Up at the top of the screen, and enter **DK73DW1215** in the promo code box on the registration screen. Access to www.Cram101.com is normally $9.95, but because you have purchased this book, your access fee is only $4.95. Sign up and stop highlighting textbooks forever.

Probability and Statistics
DeGroot and Schervish, 3rd

CONTENTS

Probability	Probability is the ratio of the number of favorable outcomes to the number of possible outcomes.
Relative frequency	In a series of observations, or trials, the relative frequency of occurrence of an event E is calculated as the number of times the event E happened over the total number of observations made. The relative frequency density of occurrence of an event is the relative frequency of E divided by the size of the bin used to classify E.
Mean	For a real-valued random variable X, the mean is the expectation of X. If the expectation does not exist, then the random variable has no mean. For a data set, the mean is just the sum of all the observations divided by the number of observations.
Frequency	Frequency is the measurement of the number of times that a repeated event occurs per unit of time.
Combinations	Combinations are un-ordered collections of unique elements. The order of the elements is not important
Statistic	A statistic is a characteristic of a sample drawn from a population.
Sample	A sample is that part of a population which is actually observed. In normal scientific practice, we demand that it is selected in such a way as to avoid presenting a biased view of the population.
Disjoint	In mathematics, two sets are said to be disjoint if they have no element in common. For example, {1, 2, 3} and {4, 5, 6} are disjoint sets.
Subset	A is a subset of a set B, if A is "contained" inside B. The relationship of one set being a subset of another is called inclusion. Every set is a subset of itself.
Union	The union of a collection of sets is the set that contains everything that belongs to any of the sets, but nothing else.
Sets	Sets are collections of objects considered as a whole. The objects of sets are called elements or members. The elements of a set can be anything: numbers, people, letters of the alphabet, other sets, and so on. Sets are conventionally denoted with capital letters, A, B, C, etc. Two sets A and B are said to be equal, written A = B, if they have the same members.
Intersection	The intersection of two sets A and B, $A \cap B$, is the set that contains all elements of A that also belong to B (or equivalently, all elements of B that also belong to A), but no other elements. The probability that two events, A and B, will both occur is $P(A \cap B) = P(A)*P(B)$.
Complement	Generally a complement of X is something that together with X makes a complete whole; that supplies what X lacks.
Empty set	The empty set is the unique set which contains no elements. In axiomatic set theory it is postulated to exist by the axiom of empty set and all finite sets are constructed from it.
Sample space	The sample space of an experiment or random trial is the set of all possible outcomes. For example, if the experiment is tossing a coin, the sample space is the set {head, tail}.
Set theory	Set theory represents collections of abstract objects. It encompasses the everyday notions, introduced in primary school, of collections of objects, and the elements of, and membership in, such collections. In most modern mathematical formalisms, set theory provides the language in which mathematical objects are described.
Elements	A set is a collection of objects considered as a whole. The objects of a set are called elements or members. The elements of a set can be anything: numbers, people, letters of the alphabet, other sets, and so on.
Natural number	In mathematics, a natural number is either a positive integer (1, 2, 3, 4, ...) or a non-negative integer (0, 1, 2, 3, 4, ...). The former definition is generally used in number theory, while the latter is preferred in set theory and computer science.
Venn diagram	A Venn diagram is an illustration used in the branch of mathematics known as set theory. They are used to show the mathematical or logical relationship between different groups of things (sets). A Venn

2

Go to **Cram101.com** for the Practice Tests for this Chapter.

diagram shows all the logical relations between the sets.

Mutually exclusive

In probability theory, events E_1, E_2, ..., E_n are said to be mutually exclusive if the occurrence of any one them automatically implies the non-occurrence of the remaining $n - 1$ events. In other words, two mutually exclusive events cannot both occur.

Range

In descriptive statistics, the range is the length of the smallest interval which contains all the data. It is calculated by subtracting the smallest observations from the greatest and provides an indication of statistical dispersion.

Probability distribution

Every random variable gives rise to a probability distribution, containing the most important information about the variable. If X is a random variable, the corresponding probability distribution assigns to the interval (a, b) the probability Pr(a ¡Ü X ¡Ü b), i.e. the probability that the variable X will take a value in the interval (a, b).

Theorem

A theorem is a proposition that has been or is to be proved on the basis of explicit assumptions.

Distribution

A distribution is a list of the values that a variable takes in a sample. It is usually a list, ordered by quantity.

Multiplication Rule

The multiplication rule states that the probability of two or more events happening is equal to the product of their individual probabilities

Permutation

The concept of a permutation expresses the idea that distinguishable objects may be arranged in various different orders. For instance, with the numbers one to six, each possible order makes a list of the numbers, without repetitions.

Sampling without replacement

Sampling without replacement means that in each successive trial of an experiment or process, the total number of possible outcomes or the mix of possible outcomes is changed by sampling. The probability of future events is thus changed.

Sampling

Sampling is that part of statistical practice concerned with the selection of individual observations intended to yield some knowledge about a population of concern, especially for the purposes of statistical inference.

Binomial coefficient

A special case of determining the number of combinations occurs when there are only two possible outcomes, such as success or failure, for each of the trials. $_nC_r = n! / r! \, (n-r)!$; or $_nC_2 = n! / 2! \, (n-2)!$ This special case is called a binomial coefficient.

Binomial

In elementary algebra, a binomial is a polynomial with two terms: the sum of two monomials. It is the simplest kind of polynomial.

Common sense

There are two general meanings to the term common sense in philosophy. One is a sense that is common to the others, and the other meaning is a sense of things that is common to humanity.

Sample	A sample is that part of a population which is actually observed. In normal scientific practice, we demand that it is selected in such a way as to avoid presenting a biased view of the population.
Sample space	The sample space of an experiment or random trial is the set of all possible outcomes. For example, if the experiment is tossing a coin, the sample space is the set {head, tail}.
Probability	Probability is the ratio of the number of favorable outcomes to the number of possible outcomes.
Placebo	A placebo is an inactive substance (pill, liquid, etc.), which is administered as if it were a therapy, but which has no therapeutic value.
Conditional probability	Conditional probability is the probability of some event A, given that some other event, B, has already occurred. Conditional probability is written P(A │ B), and is read "the probability of A, given B".
Multiplication Rule	The multiplication rule states that the probability of two or more events happening is equal to the product of their individual probabilities
Independent events	You have independent events when the occurrence of one has no effect on the probability of the occurrence of the other.
Theorem	A theorem is a proposition that has been or is to be proved on the basis of explicit assumptions.
Intersection	The intersection of two sets A and B, A ∩ B, is the set that contains all elements of A that also belong to B (or equivalently, all elements of B that also belong to A), but no other elements. The probability that two events, A and B, will both occur is P(A ∩ B) = P(A)*P(B).
Odds	In probability theory and statistics the odds in favor of an event or a proposition are the quantity $p / (1 - p)$, where p is the probability of the event or proposition. The logarithm of the odds is the logit of the probability.
Population	A population is a set of entities concerning which statistical inferences are to be drawn, often based on a random sample taken from the population.
Disjoint	In mathematics, two sets are said to be disjoint if they have no element in common. For example, {1, 2, 3} and {4, 5, 6} are disjoint sets.
Union	The union of a collection of sets is the set that contains everything that belongs to any of the sets, but nothing else.
Subset	A is a subset of a set B, if A is "contained" inside B. The relationship of one set being a subset of another is called inclusion. Every set is a subset of itself.
Conditionally independent	In probability theory, two events A and B are conditionally independent given a third event C precisely if the occurrence or non-occurrence of A and B are independent events in their conditional probability distribution given C.
Power	The power of a statistical test is the probability that the test will reject a false null hypothesis, or in other words that it will not make a Type II error. The higher the power, the greater the chance of obtaining a statistically significant result when the null hypothesis is false.
Unconditional probability	The probability of one event ignoring the occurrence or nonoccurrence of some other event is called unconditional probability.
Combinations	Combinations are un-ordered collections of unique elements. The order of the elements is not important

101

Joint probability	Joint probability is the probability of two events in conjunction. That is, it is the probability of both events together. The joint probability of A and B is written P(AB).
Random sample	A sample is a subset chosen from a population for investigation. A random sample is one chosen by a method involving an unpredictable component, in the sense that the selection of any element of the population is independent of the selection of any other element.
Frequency	Frequency is the measurement of the number of times that a repeated event occurs per unit of time.
Elements	A set is a collection of objects considered as a whole. The objects of a set are called elements or members. The elements of a set can be anything: numbers, people, letters of the alphabet, other sets, and so on.
Sets	Sets are collections of objects considered as a whole. The objects of sets are called elements or members. The elements of a set can be anything: numbers, people, letters of the alphabet, other sets, and so on. Sets are conventionally denoted with capital letters, A, B, C, etc. Two sets A and B are said to be equal, written A = B, if they have the same members.
Stochastic process	A stochastic process is a random function. In the most common applications, the domain over which the function is defined is a time interval (a stochastic process of this kind is called a time series in applications) or a region of space (a stochastic process being called a random field).
Parameter	A parameter is a characteristic of a population.
Replication	Replication is repeating the creation of a phenomenon, so that the variability associated with the phenomenon can be estimated.
Construct	A construct is a mathematical or conceptual model.

Variable	A variable is a symbol denoting a quantity or symbolic representation. In mathematics, a variable often represents an unknown quantity; in computer science, it represents a place where a quantity can be stored.
Random Variable	A variable characterized by random behavior in assuming its different possible values is a random variable. Mathematically, it is described by its probability distribution, which specifies the possible values of a random variable together with the probability associated (in an appropriate sense) with each value.
Probability	Probability is the ratio of the number of favorable outcomes to the number of possible outcomes.
Discrete random variable	A probability distribution is called discrete, if it is fully characterized by a probability mass function. Thus, the distribution of a random variable X is discrete, and X is then called a discrete random variable, if: $\sum Pr(X = \mu) = 1$, as u runs through the set of all possible values of X.
Distribution	A distribution is a list of the values that a variable takes in a sample. It is usually a list, ordered by quantity.
Sets	Sets are collections of objects considered as a whole. The objects of sets are called elements or members. The elements of a set can be anything: numbers, people, letters of the alphabet, other sets, and so on. Sets are conventionally denoted with capital letters, A, B, C, etc. Two sets A and B are said to be equal, written A = B, if they have the same members.
Mean	For a real-valued random variable X, the mean is the expectation of X. If the expectation does not exist, then the random variable has no mean. For a data set, the mean is just the sum of all the observations divided by the number of observations.
Real numbers	The real numbers are intuitively defined as numbers that are in one-to-one correspondence with the points on an infinite line—the number line.
Subset	A is a subset of a set B, if A is "contained" inside B. The relationship of one set being a subset of another is called inclusion. Every set is a subset of itself.
Uniform Distribution	In mathematics, the uniform distributions are simple probability distributions. There are two types: the discrete uniform distribution; the continuous uniform distribution.
Sample	A sample is that part of a population which is actually observed. In normal scientific practice, we demand that it is selected in such a way as to avoid presenting a biased view of the population.
Sample space	The sample space of an experiment or random trial is the set of all possible outcomes. For example, if the experiment is tossing a coin, the sample space is the set {head, tail}.
Binomial distribution	The binomial distribution is the discrete probability distribution of the number of successes in a sequence of n independent yes/no experiments, each of which yields success with probability p.
Parameter	A parameter is a characteristic of a population.
Binomial	In elementary algebra, a binomial is a polynomial with two terms: the sum of two monomials. It is the simplest kind of polynomial.
Proportional	Two quantities are called proportional if they vary in such a way that one of the quantities is a constant multiple of the other, or equivalently if they have a constant ratio.
Probability Density Function	A probability density function serves to represent a probability distribution in terms of integrals. A probability density function is non-negative everywhere and its integral from $-\ddagger$ to $+\ddagger$ is equal to 1.

Probability distribution	Every random variable gives rise to a probability distribution, containing the most important information about the variable. If X is a random variable, the corresponding probability distribution assigns to the interval (a, b) the probability $Pr(a ¡Ü X ¡Ü b)$, i.e. the probability that the variable X will take a value in the interval (a, b).
Sampling distribution	A sampling distribution is the probability distribution, under repeated sampling of the population, of a given statistic (a numerical quantity calculated from the data values in a sample).
Sampling	Sampling is that part of statistical practice concerned with the selection of individual observations intended to yield some knowledge about a population of concern, especially for the purposes of statistical inference.
Range	In descriptive statistics, the range is the length of the smallest interval which contains all the data. It is calculated by subtracting the smallest observations from the greatest and provides an indication of statistical dispersion.
Theorem	A theorem is a proposition that has been or is to be proved on the basis of explicit assumptions.
Quantiles	Quantiles are essentially points taken at regular intervals from the cumulative distribution function of a random variable. Dividing ordered data into q essentially equal-sized data subsets is the motivation for q-quantiles; the quantiles are the data values marking the boundaries between consecutive subsets.
Percentile	In descriptive statistics, a percentile is any of the 99 values that divide the sorted data into 100 equal parts, so that each part represents 1/100th (or 1%) of the sample or population.
Quartile	In descriptive statistics, a quartile is any of the three values which divide the sorted data set into four equal parts, so that each part represents 1/4th of the sample or population.
Median	The median is a number that separates the higher half of a sample, a population, or a probability distribution from the lower half. It is the middle value in a distribution, above and below which lie an equal number of values.
Generalization	Concept A is a (strict) generalization of concept B if and only if: every instance of concept B is also an instance of concept A; and there are instances of concept A which are not instances of concept B.
Joint probability	Joint probability is the probability of two events in conjunction. That is, it is the probability of both events together. The joint probability of A and B is written P(AB).
Conditional probability	Conditional probability is the probability of some event A, given that some other event, B, has already occurred. Conditional probability is written $P(A \mid B)$, and is read "the probability of A, given B".
Multiplication Rule	The multiplication rule states that the probability of two or more events happening is equal to the product of their individual probabilities
Dummy variable	A dummy variable is a notation for a place or places in an expression, into which some definite substitution may take place, or with respect to which some operation (summation or quantification, to give two examples) may take place. The idea is related to, but somewhat deeper and more complex than, that of a placeholder (a symbol that will later be replaced by some literal string), or a wildcard character that stands for an unspecified symbol.
Power	The power of a statistical test is the probability that the test will reject a false null hypothesis, or in other words that it will not make a Type II error. The higher the power, the greater the chance of obtaining a statistically significant result when the null hypothesis is false.

Go to **Cram101.com** for the Practice Tests for this Chapter.

Union	The union of a collection of sets is the set that contains everything that belongs to any of the sets, but nothing else.
Disjoint	In mathematics, two sets are said to be disjoint if they have no element in common. For example, {1, 2, 3} and {4, 5, 6} are disjoint sets.
Independent events	You have independent events when the occurrence of one has no effect on the probability of the occurrence of the other.
Constants	In mathematics and the mathematical sciences, constants are fixed, but possibly unspecified, values.
Conditionally independent	In probability theory, two events A and B are conditionally independent given a third event C precisely if the occurrence or non-occurrence of A and B are independent events in their conditional probability distribution given C.
Multivariate	In statistics, in multivariate data, each data point has more than one scalar component, and often one is concerned with correlations between the components.
Univariate	In statistics, in univariate data, each data point has only one scalar component. Or, when the statistical technique to be used contains only one dependent variable.
Random sample	A sample is a subset chosen from a population for investigation. A random sample is one chosen by a method involving an unpredictable component, in the sense that the selection of any element of the population is independent of the selection of any other element.
Reliability	Reliability has been defined in different ways by different authors. Perhaps the best way to look at reliability is the extent to which the measurements resulting from a test are the result of characteristics of those being measured.
Population	A population is a set of entities concerning which statistical inferences are to be drawn, often based on a random sample taken from the population.
Construct	A construct is a mathematical or conceptual model.
Quotient	In more abstract branches of mathematics, the word quotient is often used to describe sets, spaces, or algebraic structures whose elements are the equivalence classes of some equivalence relation on another set, space, or algebraic structure.
Linear transformation	In mathematics, a linear transformation is a function between two vector spaces that preserves the operations of vector addition and scalar multiplication. In other words, it "preserves linear combinations".
Origin	The point of intersection, where the axes meet, is called the origin normally labeled O. With the origin labeled O, we can name the x axis Ox and the y axis Oy. The x and y axes define a plane that can be referred to as the xy plane.

Go to **Cram101.com** for the Practice Tests for this Chapter.

Expected value	The expected value (or mathematical expectation) of a random variable is the sum of the probability of each possible outcome of the experiment multiplied by its payoff ("value").
Distribution	A distribution is a list of the values that a variable takes in a sample. It is usually a list, ordered by quantity.
Variable	A variable is a symbol denoting a quantity or symbolic representation. In mathematics, a variable often represents an unknown quantity; in computer science, it represents a place where a quantity can be stored.
Random Variable	A variable characterized by random behavior in assuming its different possible values is a random variable. Mathematically, it is described by its probability distribution, which specifies the possible values of a random variable together with the probability associated (in an appropriate sense) with each value.
Mean	For a real-valued random variable X, the mean is the expectation of X. If the expectation does not exist, then the random variable has no mean. For a data set, the mean is just the sum of all the observations divided by the number of observations.
Probability	Probability is the ratio of the number of favorable outcomes to the number of possible outcomes.
Origin	The point of intersection, where the axes meet, is called the origin normally labeled O. With the origin labeled O, we can name the x axis Ox and the y axis Oy. The x and y axes define a plane that can be referred to as the xy plane.
Cauchy distribution	A random variable X follows the Cauchy distribution if its probability density function is $f(x) = (1/\pi)/(1 + x^2)$. Its mode and median are zero, but the expectation value, variance and higher moments are undefined since the corresponding integrals diverge.
Probability distribution	Every random variable gives rise to a probability distribution, containing the most important information about the variable. If X is a random variable, the corresponding probability distribution assigns to the interval (a, b) the probability Pr(a ¡Ü X ¡Ü b), i.e. the probability that the variable X will take a value in the interval (a, b).
Theorem	A theorem is a proposition that has been or is to be proved on the basis of explicit assumptions.
Ordinate	The modern Cartesian coordinate system in two dimensions is commonly defined by two axes, at right angles to each other, forming a plane (an xy-plane). The horizontal axis, the abscissa, is labeled x, and the vertical axis, the ordinate, is labeled y.
Constants	In mathematics and the mathematical sciences, constants are fixed, but possibly unspecified, values.
Sampling without replacement	Sampling without replacement means that in each successive trial of an experiment or process, the total number of possible outcomes or the mix of possible outcomes is changed by sampling. The probability of future events is thus changed.
Sampling	Sampling is that part of statistical practice concerned with the selection of individual observations intended to yield some knowledge about a population of concern, especially for the purposes of statistical inference.
Binomial distribution	The binomial distribution is the discrete probability distribution of the number of successes in a sequence of n independent yes/no experiments, each of which yields success with probability p.
Parameter	A parameter is a characteristic of a population.
Binomial	In elementary algebra, a binomial is a polynomial with two terms: the sum of two monomials. It is the simplest kind of polynomial.

Go to **Cram101.com** for the Practice Tests for this Chapter.

Sample	A sample is that part of a population which is actually observed. In normal scientific practice, we demand that it is selected in such a way as to avoid presenting a biased view of the population.
Expected number	Expected number is a term used in probability theory to denote the likely number of trials to occur before a designated event happens. It is the average number that will be derived when the experiment is repeated a large number of times.
Variance	The variance of a random variable is a measure of its statistical dispersion, indicating how far from the expected value its values typically are. The variance of a real-valued random variable is its second central moment, and it also happens to be its second cumulant.
Standard deviation	The standard deviation is the most commonly used measure of statistical dispersion. Simply put, it measures how spread out the values in a data set are. The standard deviation is defined as the square root of the variance.
Deviation	A deviation is the difference between an observed value and the expected value of a variable or function.
Dispersion	In descriptive statistics, statistical dispersion is quantifiable variation of measurements of differing members of a population within the scale on which they are measured.
Construct	A construct is a mathematical or conceptual model.
Uniform Distribution	In mathematics, the uniform distributions are simple probability distributions. There are two types: the discrete uniform distribution; the continuous uniform distribution.
Discrete uniform distribution	The discrete uniform distribution is a discrete probability distribution that can be characterized by saying that all values of a finite set of possible values are equally probable.
Power	The power of a statistical test is the probability that the test will reject a false null hypothesis, or in other words that it will not make a Type II error. The higher the power, the greater the chance of obtaining a statistically significant result when the null hypothesis is false.
Median	The median is a number that separates the higher half of a sample, a population, or a probability distribution from the lower half. It is the middle value in a distribution, above and below which lie an equal number of values.
Discrete random variable	A probability distribution is called discrete, if it is fully characterized by a probability mass function. Thus, the distribution of a random variable X is discrete, and X is then called a discrete random variable, if: $\sum Pr(X = \mu) = 1$, as u runs through the set of all possible values of X.
Correlation	Correlation indicates the strength and direction of a linear relationship between two random variables. In general statistical usage, correlation refers to the departure of two variables from independence.
Covariance	Intuitively, covariance is the measure of how much two variables vary together. That is to say, the covariance becomes more positive for each pair of values which differ from their mean in the same direction, and becomes more negative with each pair of values which differ from their mean in opposite directions.
Conditionally independent	In probability theory, two events A and B are conditionally independent given a third event C precisely if the occurrence or non-occurrence of A and B are independent events in their conditional probability distribution given C.
Random sample	A sample is a subset chosen from a population for investigation. A random sample is one chosen by a method involving an unpredictable component, in the sense that the selection of

any element of the population is independent of the selection of any other element.

Sample Mean

The arithmetic mean of a set of numbers is the sum of all the members of the set divided by the number of items in the set. If the set is a statistical population, then we speak of the population mean; if of a sampling of a population, it is a sample mean.

Law of large numbers

The law of large numbers imply that the average of a random sample from a large population is likely to be close to the mean of the whole population.

Sample size

Sensitivity can be increased by using statistical controls, by increasing the reliability of measures (as in psychometric reliability), and by increasing the size of the sample. Increasing sample size is the most commonly used method for increasing statistical power.

Range

In descriptive statistics, the range is the length of the smallest interval which contains all the data. It is calculated by subtracting the smallest observations from the greatest and provides an indication of statistical dispersion.

Multivariate	In statistics, in multivariate data, each data point has more than one scalar component, and often one is concerned with correlations between the components.
Distribution	A distribution is a list of the values that a variable takes in a sample. It is usually a list, ordered by quantity.
Univariate	In statistics, in univariate data, each data point has only one scalar component. Or, when the statistical technique to be used contains only one dependent variable.
Geometric Distribution	Geometric Distribution refers to the probability of the number of times needed to do something until getting a desired result.
Binomial	In elementary algebra, a binomial is a polynomial with two terms: the sum of two monomials. It is the simplest kind of polynomial.
Bernoulli distribution	The Bernoulli distribution, named after Swiss scientist James Bernoulli, is a discrete probability distribution, which takes value 1 with success probability p and value 0 with failure probability q = 1 − p.
Binomial distribution	The binomial distribution is the discrete probability distribution of the number of successes in a sequence of n independent yes/no experiments, each of which yields success with probability p.
Variable	A variable is a symbol denoting a quantity or symbolic representation. In mathematics, a variable often represents an unknown quantity; in computer science, it represents a place where a quantity can be stored.
Random Variable	A variable characterized by random behavior in assuming its different possible values is a random variable. Mathematically, it is described by its probability distribution, which specifies the possible values of a random variable together with the probability associated (in an appropriate sense) with each value.
Parameter	A parameter is a characteristic of a population.
Probability	Probability is the ratio of the number of favorable outcomes to the number of possible outcomes.
Bernoulli trial	In the theory of probability and statistics, a Bernoulli trial is an experiment whose outcome is random and can be either of two possible outcomes, called "success" and "failure."
Population	A population is a set of entities concerning which statistical inferences are to be drawn, often based on a random sample taken from the population.
Mean	For a real-valued random variable X, the mean is the expectation of X. If the expectation does not exist, then the random variable has no mean. For a data set, the mean is just the sum of all the observations divided by the number of observations.
Mode	The mode is the value that has the largest number of observations, namely the most frequent value within a particular set of values.
Sampling without replacement	Sampling without replacement means that in each successive trial of an experiment or process, the total number of possible outcomes or the mix of possible outcomes is changed by sampling. The probability of future events is thus changed.
Sampling	Sampling is that part of statistical practice concerned with the selection of individual observations intended to yield some knowledge about a population of concern, especially for the purposes of statistical inference.
Sample	A sample is that part of a population which is actually observed. In normal scientific practice, we demand that it is selected in such a way as to avoid presenting a biased view of the population.

23

Variance	The variance of a random variable is a measure of its statistical dispersion, indicating how far from the expected value its values typically are. The variance of a real-valued random variable is its second central moment, and it also happens to be its second cumulant.
Binomial coefficient	A special case of determining the number of combinations occurs when there are only two possible outcomes, such as success or failure, for each of the trials. $_nC_r$ = n! / r! (n-r)! ; or $_nC_2$ = n! / 2! (n-2)! This special case is called a binomial coefficient.
Placebo	A placebo is an inactive substance (pill, liquid, etc.), which is administered as if it were a therapy, but which has no therapeutic value.
Poisson distribution	The Poisson distribution is a discrete probability distribution. It expresses the probability of a number of events occurring in a fixed time if these events occur with a known average rate, and are independent of the time since the last event.
Theorem	A theorem is a proposition that has been or is to be proved on the basis of explicit assumptions.
Disjoint	In mathematics, two sets are said to be disjoint if they have no element in common. For example, {1, 2, 3} and {4, 5, 6} are disjoint sets.
Expected number	Expected number is a term used in probability theory to denote the likely number of trials to occur before a designated event happens. It is the average number that will be derived when the experiment is repeated a large number of times.
Conditional probability	Conditional probability is the probability of some event A, given that some other event, B, has already occurred. Conditional probability is written $P(A \mid B)$, and is read "the probability of A, given B".
Normal distribution	The normal distribution is an extremely important probability distribution in many fields. It is a family of distributions of the same general form, differing in their location and scale parameters: the mean and standard deviation. The standard normal distribution is the normal distribution with a mean of zero and a standard deviation of one
Central limit theorem	The Central Limit Theorem states that if the sum of the variables has a finite variance, then it will be approximately normally distributed. Since many real processes yield distributions with finite variance, this explains the ubiquity of the normal distribution.
Random sample	A sample is a subset chosen from a population for investigation. A random sample is one chosen by a method involving an unpredictable component, in the sense that the selection of any element of the population is independent of the selection of any other element.
Histogram	A histogram is a graphical display of tabulated frequencies. That is, a histogram is the graphical version of a table which shows what proportion of cases fall into each of several or many specified categories. The categories are usually specified as nonoverlapping intervals of some variable. The categories (bars) must be adjacent.
Measurement	Measurement generally refers to the process of estimating or determining the ratio of a magnitude of a quantitative property or relation to a unit of the same type of quantitative property or relation.
Expected value	The expected value (or mathematical expectation) of a random variable is the sum of the probability of each possible outcome of the experiment multiplied by its payoff ("value").
Cauchy distribution	A random variable X follows the Cauchy distribution if its probability density function is $f(x) = (1/\pi)/(1 + x^2)$. Its mode and median are zero, but the expectation value, variance and higher moments are undefined since the corresponding integrals diverge.
Constants	In mathematics and the mathematical sciences, constants are fixed, but possibly unspecified, values.

Go to **Cram101.com** for the Practice Tests for this Chapter.

Standard normal distribution	The standard normal distribution is the normal distribution with a mean of zero and a standard deviation of one (the green curves in the plots to the right). It is often called the bell curve because the graph of its probability density resembles a bell.
Dummy variable	A dummy variable is a notation for a place or places in an expression, into which some definite substitution may take place, or with respect to which some operation (summation or quantification, to give two examples) may take place. The idea is related to, but somewhat deeper and more complex than, that of a placeholder (a symbol that will later be replaced by some literal string), or a wildcard character that stands for an unspecified symbol.
Statistical analysis	Statistical analysis refers to the branch of mathematics that deals with the collection, analysis, interpretation and presentation of masses of numerical data.
Standard deviation	The standard deviation is the most commonly used measure of statistical dispersion. Simply put, it measures how spread out the values in a data set are. The standard deviation is defined as the square root of the variance.
Deviation	A deviation is the difference between an observed value and the expected value of a variable or function.
Quantiles	Quantiles are essentially points taken at regular intervals from the cumulative distribution function of a random variable. Dividing ordered data into q essentially equal-sized data subsets is the motivation for q-quantiles; the quantiles are the data values marking the boundaries between consecutive subsets.
Quotient	In more abstract branches of mathematics, the word quotient is often used to describe sets, spaces, or algebraic structures whose elements are the equivalence classes of some equivalence relation on another set, space, or algebraic structure.
Sample Mean	The arithmetic mean of a set of numbers is the sum of all the members of the set divided by the number of items in the set. If the set is a statistical population, then we speak of the population mean; if of a sampling of a population, it is a sample mean.
Statistic	A statistic is a characteristic of a sample drawn from a population.
Discrete random variable	A probability distribution is called discrete, if it is fully characterized by a probability mass function. Thus, the distribution of a random variable X is discrete, and X is then called a discrete random variable, if: $\sum Pr(X = \mu) = 1$, as u runs through the set of all possible values of X.
Factorial	The factorial of a natural number n is the product of all positive integers less than and equal to n. This is written as n! and pronounced "n factorial". The notation n! was introduced by Christian Kramp in 1808.
Uniform Distribution	In mathematics, the uniform distributions are simple probability distributions. There are two types: the discrete uniform distribution; the continuous uniform distribution.
Conditionally independent	In probability theory, two events A and B are conditionally independent given a third event C precisely if the occurrence or non-occurrence of A and B are independent events in their conditional probability distribution given C.
Union	The union of a collection of sets is the set that contains everything that belongs to any of the sets, but nothing else.
Covariance	Intuitively, covariance is the measure of how much two variables vary together. That is to say, the covariance becomes more positive for each pair of values which differ from their mean in the same direction, and becomes more negative with each pair of values which differ from their mean in opposite directions.
Generalization	Concept A is a (strict) generalization of concept B if and only if: every instance of concept

	B is also an instance of concept A; and there are instances of concept A which are not instances of concept B.
Combinations	Combinations are un-ordered collections of unique elements. The order of the elements is not important
Correlation	Correlation indicates the strength and direction of a linear relationship between two random variables. In general statistical usage, correlation refers to the departure of two variables from independence.
Slope	The slope is commonly used to describe the measurement of the steepness, incline or grade of a straight line. A higher slope value indicates a steeper incline.
Proportional	Two quantities are called proportional if they vary in such a way that one of the quantities is a constant multiple of the other, or equivalently if they have a constant ratio.

Statistical Inference	Statistical inference is inference about a population from a random sample drawn from it or, more generally, about a random process from its observed behavior during a finite period of time. It includes: point estimation, interval estimation, hypothesis testing (or statistical significance testing) prediction
Mean	For a real-valued random variable X, the mean is the expectation of X. If the expectation does not exist, then the random variable has no mean. For a data set, the mean is just the sum of all the observations divided by the number of observations.
Distribution	A distribution is a list of the values that a variable takes in a sample. It is usually a list, ordered by quantity.
Probability	Probability is the ratio of the number of favorable outcomes to the number of possible outcomes.
Probability distribution	Every random variable gives rise to a probability distribution, containing the most important information about the variable. If X is a random variable, the corresponding probability distribution assigns to the interval (a, b) the probability Pr(a ¡Ü X ¡Ü b), i.e. the probability that the variable X will take a value in the interval (a, b).
Statistic	A statistic is a characteristic of a sample drawn from a population.
Parameter	A parameter is a characteristic of a population.
Sample	A sample is that part of a population which is actually observed. In normal scientific practice, we demand that it is selected in such a way as to avoid presenting a biased view of the population.
Random sample	A sample is a subset chosen from a population for investigation. A random sample is one chosen by a method involving an unpredictable component, in the sense that the selection of any element of the population is independent of the selection of any other element.
Population	A population is a set of entities concerning which statistical inferences are to be drawn, often based on a random sample taken from the population.
Statistical analysis	Statistical analysis refers to the branch of mathematics that deals with the collection, analysis, interpretation and presentation of masses of numerical data.
Variable	A variable is a symbol denoting a quantity or symbolic representation. In mathematics, a variable often represents an unknown quantity; in computer science, it represents a place where a quantity can be stored.
Random Variable	A variable characterized by random behavior in assuming its different possible values is a random variable. Mathematically, it is described by its probability distribution, which specifies the possible values of a random variable together with the probability associated (in an appropriate sense) with each value.
A priori	In statistics, a priori knowledge refers to a knowledge of the actual population, rather than that estimated by observation.
Variance	The variance of a random variable is a measure of its statistical dispersion, indicating how far from the expected value its values typically are. The variance of a real-valued random variable is its second central moment, and it also happens to be its second cumulant.
Conditionally independent	In probability theory, two events A and B are conditionally independent given a third event C precisely if the occurrence or non-occurrence of A and B are independent events in their conditional probability distribution given C.
Proportional	Two quantities are called proportional if they vary in such a way that one of the quantities is a constant multiple of the other, or equivalently if they have a constant ratio.

Bernoulli distribution	The Bernoulli distribution, named after Swiss scientist James Bernoulli, is a discrete probability distribution, which takes value 1 with success probability p and value 0 with failure probability q = 1 − p.
Sampling	Sampling is that part of statistical practice concerned with the selection of individual observations intended to yield some knowledge about a population of concern, especially for the purposes of statistical inference.
Theorem	A theorem is a proposition that has been or is to be proved on the basis of explicit assumptions.
Uniform Distribution	In mathematics, the uniform distributions are simple probability distributions. There are two types: the discrete uniform distribution; the continuous uniform distribution.
Poisson distribution	The Poisson distribution is a discrete probability distribution. It expresses the probability of a number of events occurring in a fixed time if these events occur with a known average rate, and are independent of the time since the last event.
Normal distribution	The normal distribution is an extremely important probability distribution in many fields. It is a family of distributions of the same general form, differing in their location and scale parameters: the mean and standard deviation. The standard normal distribution is the normal distribution with a mean of zero and a standard deviation of one
Histogram	A histogram is a graphical display of tabulated frequencies. That is, a histogram is the graphical version of a table which shows what proportion of cases fall into each of several or many specified categories. The categories are usually specified as nonoverlapping intervals of some variable. The categories (bars) must be adjacent.
Measurement	Measurement generally refers to the process of estimating or determining the ratio of a magnitude of a quantitative property or relation to a unit of the same type of quantitative property or relation.
Binomial distribution	The binomial distribution is the discrete probability distribution of the number of successes in a sequence of n independent yes/no experiments, each of which yields success with probability p.
Binomial	In elementary algebra, a binomial is a polynomial with two terms: the sum of two monomials. It is the simplest kind of polynomial.
Standard deviation	The standard deviation is the most commonly used measure of statistical dispersion. Simply put, it measures how spread out the values in a data set are. The standard deviation is defined as the square root of the variance.
Deviation	A deviation is the difference between an observed value and the expected value of a variable or function.
Estimator	An estimator is a function of the known sample data that is used to estimate an unknown population parameter; an estimate is the result from the actual application of the function to a particular set of data. Many different estimators are possible for any given parameter.
Median	The median is a number that separates the higher half of a sample, a population, or a probability distribution from the lower half. It is the middle value in a distribution, above and below which lie an equal number of values.
Sample size	Sensitivity can be increased by using statistical controls, by increasing the reliability of measures (as in psychometric reliability), and by increasing the size of the sample. Increasing sample size is the most commonly used method for increasing statistical power.
Law of large numbers	The law of large numbers imply that the average of a random sample from a large population is likely to be close to the mean of the whole population.

Go to **Cram101.com** for the Practice Tests for this Chapter.
And, **NEVER** highlight a book again!

Weighted average	In statistics, given a set of data, X = { x1, x2, ..., xn} and corresponding non-negative weights, W = { w1, w2, ..., wn} the weighted average, is calculated as: Mean = $\sum$w.x. / $\sum$w.
Bernoulli trial	In the theory of probability and statistics, a Bernoulli trial is an experiment whose outcome is random and can be either of two possible outcomes, called "success" and "failure."
Sample Mean	The arithmetic mean of a set of numbers is the sum of all the members of the set divided by the number of items in the set. If the set is a statistical population, then we speak of the population mean; if of a sampling of a population, it is a sample mean.
Sample variance	The variance of a random variable is a measure of its statistical dispersion, indicating how far from the expected value its values typically are. The variance of a real-valued random variable is its second central moment, and also its second cumulant. When derived from a sample rather than a population, it is a sample variance.
Cauchy distribution	A random variable X follows the Cauchy distribution if its probability density function is $f(x) = (1/\pi)/(1 + x^2)$. Its mode and median are zero, but the expectation value, variance and higher moments are undefined since the corresponding integrals diverge.
Joint probability	Joint probability is the probability of two events in conjunction. That is, it is the probability of both events together. The joint probability of A and B is written P(AB).
Randomization	Randomization involves randomly allocating the experimental units across the treatment groups. Thus, if the experiment compares a new drug against a standard drug used as a control, the patients should be allocated to new drug or control by a random process.
Geometric Distribution	Geometric Distribution refers to the probability of the number of times needed to do something until getting a desired result.

Sampling distribution	A sampling distribution is the probability distribution, under repeated sampling of the population, of a given statistic (a numerical quantity calculated from the data values in a sample).
Statistic	A statistic is a characteristic of a sample drawn from a population.
Sampling	Sampling is that part of statistical practice concerned with the selection of individual observations intended to yield some knowledge about a population of concern, especially for the purposes of statistical inference.
Distribution	A distribution is a list of the values that a variable takes in a sample. It is usually a list, ordered by quantity.
Parameter	A parameter is a characteristic of a population.
Variable	A variable is a symbol denoting a quantity or symbolic representation. In mathematics, a variable often represents an unknown quantity; in computer science, it represents a place where a quantity can be stored.
Random Variable	A variable characterized by random behavior in assuming its different possible values is a random variable. Mathematically, it is described by its probability distribution, which specifies the possible values of a random variable together with the probability associated (in an appropriate sense) with each value.
Probability	Probability is the ratio of the number of favorable outcomes to the number of possible outcomes.
Binomial distribution	The binomial distribution is the discrete probability distribution of the number of successes in a sequence of n independent yes/no experiments, each of which yields success with probability p.
Binomial	In elementary algebra, a binomial is a polynomial with two terms: the sum of two monomials. It is the simplest kind of polynomial.
Sample	A sample is that part of a population which is actually observed. In normal scientific practice, we demand that it is selected in such a way as to avoid presenting a biased view of the population.
Random sample	A sample is a subset chosen from a population for investigation. A random sample is one chosen by a method involving an unpredictable component, in the sense that the selection of any element of the population is independent of the selection of any other element.
Estimator	An estimator is a function of the known sample data that is used to estimate an unknown population parameter; an estimate is the result from the actual application of the function to a particular set of data. Many different estimators are possible for any given parameter.
Sample size	Sensitivity can be increased by using statistical controls, by increasing the reliability of measures (as in psychometric reliability), and by increasing the size of the sample. Increasing sample size is the most commonly used method for increasing statistical power.
Degrees of freedom	In fitting statistical models to data, the vectors of residuals are often constrained to lie in a space of smaller dimension than the number of components in the vector. That smaller dimension is the number of degrees of freedom for error.
Theorem	A theorem is a proposition that has been or is to be proved on the basis of explicit assumptions.
Normal distribution	The normal distribution is an extremely important probability distribution in many fields. It is a family of distributions of the same general form, differing in their location and scale parameters: the mean and standard deviation. The standard normal distribution is the normal distribution with a mean of zero and a standard deviation of one

Standard normal distribution	The standard normal distribution is the normal distribution with a mean of zero and a standard deviation of one (the green curves in the plots to the right). It is often called the bell curve because the graph of its probability density resembles a bell.
Variance	The variance of a random variable is a measure of its statistical dispersion, indicating how far from the expected value its values typically are. The variance of a real-valued random variable is its second central moment, and it also happens to be its second cumulant.
Mean	For a real-valued random variable X, the mean is the expectation of X. If the expectation does not exist, then the random variable has no mean. For a data set, the mean is just the sum of all the observations divided by the number of observations.
Mode	The mode is the value that has the largest number of observations, namely the most frequent value within a particular set of values.
Median	The median is a number that separates the higher half of a sample, a population, or a probability distribution from the lower half. It is the middle value in a distribution, above and below which lie an equal number of values.
Sample variance	The variance of a random variable is a measure of its statistical dispersion, indicating how far from the expected value its values typically are. The variance of a real-valued random variable is its second central moment, and also its second cumulant. When derived from a sample rather than a population, it is a sample variance.
Sample Mean	The arithmetic mean of a set of numbers is the sum of all the members of the set divided by the number of items in the set. If the set is a statistical population, then we speak of the population mean; if of a sampling of a population, it is a sample mean.
Orthogonal	In mathematics, orthogonal is synonymous with perpendicular when used as a simple adjective that is not part of any longer phrase with a standard definition. It means at right angles.
Linear transformation	In mathematics, a linear transformation is a function between two vector spaces that preserves the operations of vector addition and scalar multiplication. In other words, it "preserves linear combinations".
Standard deviation	The standard deviation is the most commonly used measure of statistical dispersion. Simply put, it measures how spread out the values in a data set are. The standard deviation is defined as the square root of the variance.
Deviation	A deviation is the difference between an observed value and the expected value of a variable or function.
Construct	A construct is a mathematical or conceptual model.
Statistical Inference	Statistical inference is inference about a population from a random sample drawn from it or, more generally, about a random process from its observed behavior during a finite period of time. It includes: point estimation, interval estimation, hypothesis testing (or statistical significance testing) prediction
Cauchy distribution	A random variable X follows the Cauchy distribution if its probability density function is $f(x) = (1/\pi)/(1 + x^2)$. Its mode and median are zero, but the expectation value, variance and higher moments are undefined since the corresponding integrals diverge.
Confidence Interval	A confidence interval is an interval between two numbers, where there is a certain specified level of confidence that a population parameter lies.
Measurement	Measurement generally refers to the process of estimating or determining the ratio of a magnitude of a quantitative property or relation to a unit of the same type of quantitative property or relation.
Probability	Every random variable gives rise to a probability distribution, containing the most important

distribution	information about the variable. If X is a random variable, the corresponding probability distribution assigns to the interval (a, b) the probability Pr(a ¡Ü X ¡Ü b), i.e. the probability that the variable X will take a value in the interval (a, b).
Interval estimate	An Interval estimate is a range of values estimated to include the parameter.
Proportional	Two quantities are called proportional if they vary in such a way that one of the quantities is a constant multiple of the other, or equivalently if they have a constant ratio.
Weighted average	In statistics, given a set of data, X = { x1, x2, ..., xn} and corresponding non-negative weights, W = { w1, w2, ..., wn} the weighted average, is calculated as: Mean = $\sum$w.x. / $\sum$w.
Bias	A bias is a prejudice in a general or specific sense, usually in the sense for having a predilection to one particular point of view or ideological perspective. However, one is generally only said to be biased if one's powers of judgement are influenced by the biases one holds, to the extent that one's views could not be taken as being neutral or objective, but instead as subjective.
Poisson distribution	The Poisson distribution is a discrete probability distribution. It expresses the probability of a number of events occurring in a fixed time if these events occur with a known average rate, and are independent of the time since the last event.
Geometric Distribution	Geometric Distribution refers to the probability of the number of times needed to do something until getting a desired result.
Bernoulli distribution	The Bernoulli distribution, named after Swiss scientist James Bernoulli, is a discrete probability distribution, which takes value 1 with success probability p and value 0 with failure probability q = 1 − p.
Constants	In mathematics and the mathematical sciences, constants are fixed, but possibly unspecified, values.
Subset	A is a subset of a set B, if A is "contained" inside B. The relationship of one set being a subset of another is called inclusion. Every set is a subset of itself.
Uniform Distribution	In mathematics, the uniform distributions are simple probability distributions. There are two types: the discrete uniform distribution; the continuous uniform distribution.

Go to **Cram101.com** for the Practice Tests for this Chapter.

Hypothesis testing	Hypothesis testing is an algorithm to state the alternative which minimizes certain risks.
Parameter	A parameter is a characteristic of a population.
Subset	A is a subset of a set B, if A is "contained" inside B. The relationship of one set being a subset of another is called inclusion. Every set is a subset of itself.
Hypothesis	A hypothesis is a proposed explanation for a phenomenon. A scientific hypothesis must be testable and based on previous observations or extensions of scientific theories.
Complement	Generally a complement of X is something that together with X makes a complete whole; that supplies what X lacks.
Alternative hypothesis	The alternate hypothesis, or alternative hypothesis, together with the null hypothesis are the two rival hypothesis whose likelihoods are compared by a statistical hypothesis test. Usually the alternate hypothesis is the possibility that an observed effect is genuine and the null hypothesis is the rival possibility that it has resulted from random chance.
Null hypothesis	A null hypothesis, H_o, is a hypothesis set up to be nullified or refuted in order to support an alternative hypothesis. When used, the null hypothesis is presumed true until statistical evidence in the form of a hypothesis test indicates otherwise.
Measurement	Measurement generally refers to the process of estimating or determining the ratio of a magnitude of a quantitative property or relation to a unit of the same type of quantitative property or relation.
Variance	The variance of a random variable is a measure of its statistical dispersion, indicating how far from the expected value its values typically are. The variance of a real-valued random variable is its second central moment, and it also happens to be its second cumulant.
Mean	For a real-valued random variable X, the mean is the expectation of X. If the expectation does not exist, then the random variable has no mean. For a data set, the mean is just the sum of all the observations divided by the number of observations.
Composite Hypothesis	A statistical hypothesis which does not completely specify the distribution of a random variable is referred to as a composite hypothesis.
Simple Hypothesis	A simple hypothesis is a hypothesis which specifies the population distribution completely.
Distribution	A distribution is a list of the values that a variable takes in a sample. It is usually a list, ordered by quantity.
Sample	A sample is that part of a population which is actually observed. In normal scientific practice, we demand that it is selected in such a way as to avoid presenting a biased view of the population.
Sample space	The sample space of an experiment or random trial is the set of all possible outcomes. For example, if the experiment is tossing a coin, the sample space is the set {head, tail}.
Statistic	A statistic is a characteristic of a sample drawn from a population.
Test Statistic	A test statistic is a summary value (often a summary statistic) of a data set that is compared with a statistical distribution to determine whether the data set differs from that expected under a null hypothesis.
Normal distribution	The normal distribution is an extremely important probability distribution in many fields. It is a family of distributions of the same general form, differing in their location and scale parameters: the mean and standard deviation. The standard normal distribution is the normal distribution with a mean of zero and a standard deviation of one

Go to **Cram101.com** for the Practice Tests for this Chapter.

Power	The power of a statistical test is the probability that the test will reject a false null hypothesis, or in other words that it will not make a Type II error. The higher the power, the greater the chance of obtaining a statistically significant result when the null hypothesis is false.
Probability	Probability is the ratio of the number of favorable outcomes to the number of possible outcomes.
Probability of error	The probability of error is for a Type I error, shown as α (alpha) and is known as the size of the test and is 1 minus the specificity of the test. For a Type II error, it is shown as β (beta) and is 1 minus the power or 1 minus the sensitivity of the test.
Uniform Distribution	In mathematics, the uniform distributions are simple probability distributions. There are two types: the discrete uniform distribution; the continuous uniform distribution.
Binomial distribution	The binomial distribution is the discrete probability distribution of the number of successes in a sequence of n independent yes/no experiments, each of which yields success with probability p.
Binomial	In elementary algebra, a binomial is a polynomial with two terms: the sum of two monomials. It is the simplest kind of polynomial.
Confidence Interval	A confidence interval is an interval between two numbers, where there is a certain specified level of confidence that a population parameter lies.
Sets	Sets are collections of objects considered as a whole. The objects of sets are called elements or members. The elements of a set can be anything: numbers, people, letters of the alphabet, other sets, and so on. Sets are conventionally denoted with capital letters, A, B, C, etc. Two sets A and B are said to be equal, written A = B, if they have the same members.
Construct	A construct is a mathematical or conceptual model.
Degrees of freedom	In fitting statistical models to data, the vectors of residuals are often constrained to lie in a space of smaller dimension than the number of components in the vector. That smaller dimension is the number of degrees of freedom for error.
Random sample	A sample is a subset chosen from a population for investigation. A random sample is one chosen by a method involving an unpredictable component, in the sense that the selection of any element of the population is independent of the selection of any other element.
Variable	A variable is a symbol denoting a quantity or symbolic representation. In mathematics, a variable often represents an unknown quantity; in computer science, it represents a place where a quantity can be stored.
Random Variable	A variable characterized by random behavior in assuming its different possible values is a random variable. Mathematically, it is described by its probability distribution, which specifies the possible values of a random variable together with the probability associated (in an appropriate sense) with each value.
Hypothesis test	One may be faced with the problem of making a definite decision with respect to an uncertain hypothesis which is known only through its observable consequences. A statistical hypothesis test, or more briefly, hypothesis test, is an algorithm to state the alternative which minimizes certain risks.
Significance level	The significance level of a test is the maximum probability of accidentally rejecting a true null hypothesis (a decision known as a Type I error). The significance of a result is also called its p-value; the smaller the p-value, the more significant the result is said to be.
Theorem	A theorem is a proposition that has been or is to be proved on the basis of explicit assumptions.

Go to **Cram101.com** for the Practice Tests for this Chapter.

Sampling	Sampling is that part of statistical practice concerned with the selection of individual observations intended to yield some knowledge about a population of concern, especially for the purposes of statistical inference.
Sample Mean	The arithmetic mean of a set of numbers is the sum of all the members of the set divided by the number of items in the set. If the set is a statistical population, then we speak of the population mean; if of a sampling of a population, it is a sample mean.
Bernoulli distribution	The Bernoulli distribution, named after Swiss scientist James Bernoulli, is a discrete probability distribution, which takes value 1 with success probability p and value 0 with failure probability q = 1 − p.
Randomization	Randomization involves randomly allocating the experimental units across the treatment groups. Thus, if the experiment compares a new drug against a standard drug used as a control, the patients should be allocated to new drug or control by a random process.
Standard normal distribution	The standard normal distribution is the normal distribution with a mean of zero and a standard deviation of one (the green curves in the plots to the right). It is often called the bell curve because the graph of its probability density resembles a bell.
Sample size	Sensitivity can be increased by using statistical controls, by increasing the reliability of measures (as in psychometric reliability), and by increasing the size of the sample. Increasing sample size is the most commonly used method for increasing statistical power.
Constants	In mathematics and the mathematical sciences, constants are fixed, but possibly unspecified, values.
Quantiles	Quantiles are essentially points taken at regular intervals from the cumulative distribution function of a random variable. Dividing ordered data into q essentially equal-sized data subsets is the motivation for q-quantiles; the quantiles are the data values marking the boundaries between consecutive subsets.
Placebo	A placebo is an inactive substance (pill, liquid, etc.), which is administered as if it were a therapy, but which has no therapeutic value.
Estimator	An estimator is a function of the known sample data that is used to estimate an unknown population parameter; an estimate is the result from the actual application of the function to a particular set of data. Many different estimators are possible for any given parameter.
Median	The median is a number that separates the higher half of a sample, a population, or a probability distribution from the lower half. It is the middle value in a distribution, above and below which lie an equal number of values.
Sampling distribution	A sampling distribution is the probability distribution, under repeated sampling of the population, of a given statistic (a numerical quantity calculated from the data values in a sample).
Expected value	The expected value (or mathematical expectation) of a random variable is the sum of the probability of each possible outcome of the experiment multiplied by its payoff ("value").
Proportional	Two quantities are called proportional if they vary in such a way that one of the quantities is a constant multiple of the other, or equivalently if they have a constant ratio.
Deviation	A deviation is the difference between an observed value and the expected value of a variable or function.
Type II error	In statistics, a false negative, also called a Type II error or miss, exists when a test incorrectly reports that a result was not detected, when it was really present.
Poisson distribution	The Poisson distribution is a discrete probability distribution. It expresses the probability of a number of events occurring in a fixed time if these events occur with a known average

Go to **Cram101.com** for the Practice Tests for this Chapter.

rate, and are independent of the time since the last event.

Distribution	A distribution is a list of the values that a variable takes in a sample. It is usually a list, ordered by quantity.
Alternative hypothesis	The alternate hypothesis, or alternative hypothesis, together with the null hypothesis are the two rival hypothesis whose likelihoods are compared by a statistical hypothesis test. Usually the alternate hypothesis is the possibility that an observed effect is genuine and the null hypothesis is the rival possibility that it has resulted from random chance.
Null hypothesis	A null hypothesis, H_o, is a hypothesis set up to be nullified or refuted in order to support an alternative hypothesis. When used, the null hypothesis is presumed true until statistical evidence in the form of a hypothesis test indicates otherwise.
Hypothesis	A hypothesis is a proposed explanation for a phenomenon. A scientific hypothesis must be testable and based on previous observations or extensions of scientific theories.
Parameter	A parameter is a characteristic of a population.
Probability	Probability is the ratio of the number of favorable outcomes to the number of possible outcomes.
Degrees of freedom	In fitting statistical models to data, the vectors of residuals are often constrained to lie in a space of smaller dimension than the number of components in the vector. That smaller dimension is the number of degrees of freedom for error.
Uniform Distribution	In mathematics, the uniform distributions are simple probability distributions. There are two types: the discrete uniform distribution; the continuous uniform distribution.
Disjoint	In mathematics, two sets are said to be disjoint if they have no element in common. For example, {1, 2, 3} and {4, 5, 6} are disjoint sets.
Expected number	Expected number is a term used in probability theory to denote the likely number of trials to occur before a designated event happens. It is the average number that will be derived when the experiment is repeated a large number of times.
Sample	A sample is that part of a population which is actually observed. In normal scientific practice, we demand that it is selected in such a way as to avoid presenting a biased view of the population.
Variable	A variable is a symbol denoting a quantity or symbolic representation. In mathematics, a variable often represents an unknown quantity; in computer science, it represents a place where a quantity can be stored.
Random Variable	A variable characterized by random behavior in assuming its different possible values is a random variable. Mathematically, it is described by its probability distribution, which specifies the possible values of a random variable together with the probability associated (in an appropriate sense) with each value.
Normal distribution	The normal distribution is an extremely important probability distribution in many fields. It is a family of distributions of the same general form, differing in their location and scale parameters: the mean and standard deviation. The standard normal distribution is the normal distribution with a mean of zero and a standard deviation of one
Variance	The variance of a random variable is a measure of its statistical dispersion, indicating how far from the expected value its values typically are. The variance of a real-valued random variable is its second central moment, and it also happens to be its second cumulant.
Mean	For a real-valued random variable X, the mean is the expectation of X. If the expectation does not exist, then the random variable has no mean. For a data set, the mean is just the sum of all the observations divided by the number of observations.
Power	The power of a statistical test is the probability that the test will reject a false null

Go to **Cram101.com** for the Practice Tests for this Chapter.

hypothesis, or in other words that it will not make a Type II error. The higher the power, the greater the chance of obtaining a statistically significant result when the null hypothesis is false.

Deviation	A deviation is the difference between an observed value and the expected value of a variable or function.
Statistic	A statistic is a characteristic of a sample drawn from a population.
Random sample	A sample is a subset chosen from a population for investigation. A random sample is one chosen by a method involving an unpredictable component, in the sense that the selection of any element of the population is independent of the selection of any other element.
Central limit theorem	The Central Limit Theorem states that if the sum of the variables has a finite variance, then it will be approximately normally distributed. Since many real processes yield distributions with finite variance, this explains the ubiquity of the normal distribution.
Theorem	A theorem is a proposition that has been or is to be proved on the basis of explicit assumptions.
Standard normal distribution	The standard normal distribution is the normal distribution with a mean of zero and a standard deviation of one (the green curves in the plots to the right). It is often called the bell curve because the graph of its probability density resembles a bell.
Subset	A is a subset of a set B, if A is "contained" inside B. The relationship of one set being a subset of another is called inclusion. Every set is a subset of itself.
Sample size	Sensitivity can be increased by using statistical controls, by increasing the reliability of measures (as in psychometric reliability), and by increasing the size of the sample. Increasing sample size is the most commonly used method for increasing statistical power.
Sample variance	The variance of a random variable is a measure of its statistical dispersion, indicating how far from the expected value its values typically are. The variance of a real-valued random variable is its second central moment, and also its second cumulant. When derived from a sample rather than a population, it is a sample variance.
Sample Mean	The arithmetic mean of a set of numbers is the sum of all the members of the set divided by the number of items in the set. If the set is a statistical population, then we speak of the population mean; if of a sampling of a population, it is a sample mean.
Poisson distribution	The Poisson distribution is a discrete probability distribution. It expresses the probability of a number of events occurring in a fixed time if these events occur with a known average rate, and are independent of the time since the last event.
Range	In descriptive statistics, the range is the length of the smallest interval which contains all the data. It is calculated by subtracting the smallest observations from the greatest and provides an indication of statistical dispersion.
Marginal probability	Marginal probability is the probability of one event, regardless of the other event. Marginal probability is obtained by summing (or integrating, more generally) the joint probability over the unrequired event. This is called marginalization. The marginal probability of A is written P(A), and the marginal probability of B is written P(B).
Population	A population is a set of entities concerning which statistical inferences are to be drawn, often based on a random sample taken from the population.
Discrete random variable	A probability distribution is called discrete, if it is fully characterized by a probability mass function. Thus, the distribution of a random variable X is discrete, and X is then called a discrete random variable, if: $\sum Pr(X = \mu) = 1$, as u runs through the set of all possible values of X.

Go to **Cram101.com** for the Practice Tests for this Chapter.

Estimator	An estimator is a function of the known sample data that is used to estimate an unknown population parameter; an estimate is the result from the actual application of the function to a particular set of data. Many different estimators are possible for any given parameter.
Probability distribution	Every random variable gives rise to a probability distribution, containing the most important information about the variable. If X is a random variable, the corresponding probability distribution assigns to the interval (a, b) the probability Pr(a ¡Ü X ¡Ü b), i.e. the probability that the variable X will take a value in the interval (a, b).
Placebo	A placebo is an inactive substance (pill, liquid, etc.), which is administered as if it were a therapy, but which has no therapeutic value.
Measurement	Measurement generally refers to the process of estimating or determining the ratio of a magnitude of a quantitative property or relation to a unit of the same type of quantitative property or relation.
Cauchy distribution	A random variable X follows the Cauchy distribution if its probability density function is $f(x) = (1/\pi)/(1 + x^2)$. Its mode and median are zero, but the expectation value, variance and higher moments are undefined since the corresponding integrals diverge.
Median	The median is a number that separates the higher half of a sample, a population, or a probability distribution from the lower half. It is the middle value in a distribution, above and below which lie an equal number of values.
Weighted average	In statistics, given a set of data, X = { x1, x2, ..., xn} and corresponding non-negative weights, W = { w1, w2, ..., wn} the weighted average, is calculated as: Mean = $\sum w.x. / \sum w$.
Standard deviation	The standard deviation is the most commonly used measure of statistical dispersion. Simply put, it measures how spread out the values in a data set are. The standard deviation is defined as the square root of the variance.
Interquartile range	In descriptive statistics, the interquartile range (IQR) is the difference between the third and first quartiles and is a measure of statistical dispersion. The interquartile range is a more stable statistic than the range, and is often preferred to that statistic.
Quantiles	Quantiles are essentially points taken at regular intervals from the cumulative distribution function of a random variable. Dividing ordered data into q essentially equal-sized data subsets is the motivation for q-quantiles; the quantiles are the data values marking the boundaries between consecutive subsets.
Proportional	Two quantities are called proportional if they vary in such a way that one of the quantities is a constant multiple of the other, or equivalently if they have a constant ratio.
Outlier	In statistics, an outlier is a single observation "far away" from the rest of the data. In most samplings of data, some data points will be further away from their expected values than what is deemed reasonable. This can be due to systematic error or faults in the theory that generated the expected values.
Sign Test	The sign test can be used to test the hypothesis that there is "no difference" between two continuous distributions X and Y. Formally: Let p = P(X > Y), and then test the null hypothesis Ho: p = 0.50. This hypothesis implies that given a random pair of measurements (xi, yi), then both xi and yi are equally likely to be larger than the other.
Confidence Interval	A confidence interval is an interval between two numbers, where there is a certain specified level of confidence that a population parameter lies.
Construct	A construct is a mathematical or conceptual model.

Go to **Cram101.com** for the Practice Tests for this Chapter.

Variable	A variable is a symbol denoting a quantity or symbolic representation. In mathematics, a variable often represents an unknown quantity; in computer science, it represents a place where a quantity can be stored.
Sample	A sample is that part of a population which is actually observed. In normal scientific practice, we demand that it is selected in such a way as to avoid presenting a biased view of the population.
Least Squares	Least squares is a mathematical optimization technique which, when given a series of measured data, attempts to find a function which closely approximates the data (a "best fit"). It attempts to minimize the sum of the squares of the ordinate differences (called residuals) between points generated by the function and corresponding points in the data.
Deviation	A deviation is the difference between an observed value and the expected value of a variable or function.
Range	In descriptive statistics, the range is the length of the smallest interval which contains all the data. It is calculated by subtracting the smallest observations from the greatest and provides an indication of statistical dispersion.
Random Variable	A variable characterized by random behavior in assuming its different possible values is a random variable. Mathematically, it is described by its probability distribution, which specifies the possible values of a random variable together with the probability associated (in an appropriate sense) with each value.
Distribution	A distribution is a list of the values that a variable takes in a sample. It is usually a list, ordered by quantity.
Regression coefficient	The regression coefficient is the slope of the straight line that most closely relates two correlated variables.
Parameter	A parameter is a characteristic of a population.
Linear regression	In statistics, linear regression is a method of estimating the conditional expected value of one variable y given the values of some other variable or variables x. The variable of interest, y, is conventionally called the "response variable".
Variance	The variance of a random variable is a measure of its statistical dispersion, indicating how far from the expected value its values typically are. The variance of a real-valued random variable is its second central moment, and it also happens to be its second cumulant.
Estimator	An estimator is a function of the known sample data that is used to estimate an unknown population parameter; an estimate is the result from the actual application of the function to a particular set of data. Many different estimators are possible for any given parameter.
Normal distribution	The normal distribution is an extremely important probability distribution in many fields. It is a family of distributions of the same general form, differing in their location and scale parameters: the mean and standard deviation. The standard normal distribution is the normal distribution with a mean of zero and a standard deviation of one
Mean	For a real-valued random variable X, the mean is the expectation of X. If the expectation does not exist, then the random variable has no mean. For a data set, the mean is just the sum of all the observations divided by the number of observations.
Covariance	Intuitively, covariance is the measure of how much two variables vary together. That is to say, the covariance becomes more positive for each pair of values which differ from their mean in the same direction, and becomes more negative with each pair of values which differ from their mean in opposite directions.
Population	A population is a set of entities concerning which statistical inferences are to be drawn,

often based on a random sample taken from the population.

Statistic	A statistic is a characteristic of a sample drawn from a population.
Theorem	A theorem is a proposition that has been or is to be proved on the basis of explicit assumptions.
Statistical Inference	Statistical inference is inference about a population from a random sample drawn from it or, more generally, about a random process from its observed behavior during a finite period of time. It includes: point estimation, interval estimation, hypothesis testing (or statistical significance testing) prediction
Standard normal distribution	The standard normal distribution is the normal distribution with a mean of zero and a standard deviation of one (the green curves in the plots to the right). It is often called the bell curve because the graph of its probability density resembles a bell.
Combinations	Combinations are un-ordered collections of unique elements. The order of the elements is not important
Orthogonal	In mathematics, orthogonal is synonymous with perpendicular when used as a simple adjective that is not part of any longer phrase with a standard definition. It means at right angles.
Confidence Interval	A confidence interval is an interval between two numbers, where there is a certain specified level of confidence that a population parameter lies.
Test Statistic	A test statistic is a summary value (often a summary statistic) of a data set that is compared with a statistical distribution to determine whether the data set differs from that expected under a null hypothesis.
Degrees of freedom	In fitting statistical models to data, the vectors of residuals are often constrained to lie in a space of smaller dimension than the number of components in the vector. That smaller dimension is the number of degrees of freedom for error.
Prediction interval	In statistics, a prediction interval bears the same relationship to a future observation that a confidence interval bears to an unobservable population parameter.
Probability	Probability is the ratio of the number of favorable outcomes to the number of possible outcomes.
Residual	Error is a misnomer; an error is the amount by which an observation differs from its expected value; the latter being based on the whole population from which the statistical unit was chosen randomly. A residual, on the other hand, is an observable estimate of the unobservable error.
Outlier	In statistics, an outlier is a single observation "far away" from the rest of the data. In most samplings of data, some data points will be further away from their expected values than what is deemed reasonable. This can be due to systematic error or faults in the theory that generated the expected values.
Regression analysis	Regression analysis is any statistical method where the mean of one or more random variables is predicted conditioned on other (measured) random variables.
Sets	Sets are collections of objects considered as a whole. The objects of sets are called elements or members. The elements of a set can be anything: numbers, people, letters of the alphabet, other sets, and so on. Sets are conventionally denoted with capital letters, A, B, C, etc. Two sets A and B are said to be equal, written A = B, if they have the same members.
Quantiles	Quantiles are essentially points taken at regular intervals from the cumulative distribution function of a random variable. Dividing ordered data into q essentially equal-sized data subsets is the motivation for q-quantiles; the quantiles are the data values marking the boundaries between consecutive subsets.

Go to **Cram101.com** for the Practice Tests for this Chapter.
And, **NEVER** highlight a book again!

Null hypothesis	A null hypothesis, H_0, is a hypothesis set up to be nullified or refuted in order to support an alternative hypothesis. When used, the null hypothesis is presumed true until statistical evidence in the form of a hypothesis test indicates otherwise.
Hypothesis	A hypothesis is a proposed explanation for a phenomenon. A scientific hypothesis must be testable and based on previous observations or extensions of scientific theories.
Construct	A construct is a mathematical or conceptual model.
Confidence limits	Confidence limits form the upper and lower bounds of a confidence interval.
Slope	The slope is commonly used to describe the measurement of the steepness, incline or grade of a straight line. A higher slope value indicates a steeper incline.
Bayesian inference	Bayesian inference is statistical inference in which evidence or observations are used to update or to newly infer the probability that a hypothesis may be true.
Sampling distribution	A sampling distribution is the probability distribution, under repeated sampling of the population, of a given statistic (a numerical quantity calculated from the data values in a sample).
Sampling	Sampling is that part of statistical practice concerned with the selection of individual observations intended to yield some knowledge about a population of concern, especially for the purposes of statistical inference.
Proportional	Two quantities are called proportional if they vary in such a way that one of the quantities is a constant multiple of the other, or equivalently if they have a constant ratio.
Intercept	An intercept is the coordinate of the point at which a curve cuts an axis. For example, an x-intercept or a y-intercept.
Multiple regression	A multiple regression is a linear regression with more than one covariate (predictor variable). It can be viewed as a simple case of canonical correlation. An equation used to predict a dependent variable, y from two independents, u and v is: $y = \beta_0 + \beta_1 u + \beta_2 v + \beta_3 u2 + \beta_4 uv + \beta_5 v2$
Predictor variable	The predictor variable is manipulated by the experimenter. By attempting to isolate all other factors, one can determine the influence of the independent variable on the dependent variable.
Multiple linear regression	Multiple linear regression extends linear regression to functions of two or more variables, for example: $Z = aX + bY + c + \epsilon$.. Here X and Y are explanatory variables. The values of the parameters a, b and c are estimated by the method of least squares, that minimize the sum of squares of the residuals
Generalization	Concept A is a (strict) generalization of concept B if and only if: every instance of concept B is also an instance of concept A; and there are instances of concept A which are not instances of concept B.
Elements	A set is a collection of objects considered as a whole. The objects of a set are called elements or members. The elements of a set can be anything: numbers, people, letters of the alphabet, other sets, and so on.
Power	The power of a statistical test is the probability that the test will reject a false null hypothesis, or in other words that it will not make a Type II error. The higher the power, the greater the chance of obtaining a statistically significant result when the null hypothesis is false.
Sample size	Sensitivity can be increased by using statistical controls, by increasing the reliability of measures (as in psychometric reliability), and by increasing the size of the sample.

Increasing sample size is the most commonly used method for increasing statistical power.

Analysis of variance	Analysis of variance (ANOVA) is a collection of statistical models and their associated procedures which compare means by splitting the overall observed variance into different parts.
Alternative hypothesis	The alternate hypothesis, or alternative hypothesis, together with the null hypothesis are the two rival hypothesis whose likelihoods are compared by a statistical hypothesis test. Usually the alternate hypothesis is the possibility that an observed effect is genuine and the null hypothesis is the rival possibility that it has resulted from random chance.
Sum of squares	Sum of squares is a concept that permeates much of inferential statistics and descriptive statistics. More properly, it is "the sum of the squared deviations". Mathematically, it is an unscaled, or unadjusted measure of dispersion. When scaled for the number of degrees of freedom, it becomes the variance, the sum of squares per degree of freedom.
Subset	A is a subset of a set B, if A is "contained" inside B. The relationship of one set being a subset of another is called inclusion. Every set is a subset of itself.
Sample variance	The variance of a random variable is a measure of its statistical dispersion, indicating how far from the expected value its values typically are. The variance of a real-valued random variable is its second central moment, and also its second cumulant. When derived from a sample rather than a population, it is a sample variance.
Sample Mean	The arithmetic mean of a set of numbers is the sum of all the members of the set divided by the number of items in the set. If the set is a statistical population, then we speak of the population mean; if of a sampling of a population, it is a sample mean.
Expected value	The expected value (or mathematical expectation) of a random variable is the sum of the probability of each possible outcome of the experiment multiplied by its payoff ("value").
Measurement	Measurement generally refers to the process of estimating or determining the ratio of a magnitude of a quantitative property or relation to a unit of the same type of quantitative property or relation.
Interaction	Interaction is a kind of action which occurs as two or more objects have an effect upon one another. The idea of a two-way effect is essential in the concept of interaction instead of a one-way causal effect.
Replication	Replication is repeating the creation of a phenomenon, so that the variability associated with the phenomenon can be estimated.
Main effect	The main effect is the direct effect that a each independent variable has on the dependent variable without regard to the possibility of interactions.

Go to **Cram101.com** for the Practice Tests for this Chapter.

Quantiles	Quantiles are essentially points taken at regular intervals from the cumulative distribution function of a random variable. Dividing ordered data into q essentially equal-sized data subsets is the motivation for q-quantiles; the quantiles are the data values marking the boundaries between consecutive subsets.
Mean	For a real-valued random variable X, the mean is the expectation of X. If the expectation does not exist, then the random variable has no mean. For a data set, the mean is just the sum of all the observations divided by the number of observations.
Distribution	A distribution is a list of the values that a variable takes in a sample. It is usually a list, ordered by quantity.
Estimator	An estimator is a function of the known sample data that is used to estimate an unknown population parameter; an estimate is the result from the actual application of the function to a particular set of data. Many different estimators are possible for any given parameter.
Parameter	A parameter is a characteristic of a population.
Median	The median is a number that separates the higher half of a sample, a population, or a probability distribution from the lower half. It is the middle value in a distribution, above and below which lie an equal number of values.
Law of large numbers	The law of large numbers imply that the average of a random sample from a large population is likely to be close to the mean of the whole population.
Variable	A variable is a symbol denoting a quantity or symbolic representation. In mathematics, a variable often represents an unknown quantity; in computer science, it represents a place where a quantity can be stored.
Random Variable	A variable characterized by random behavior in assuming its different possible values is a random variable. Mathematically, it is described by its probability distribution, which specifies the possible values of a random variable together with the probability associated (in an appropriate sense) with each value.
Construct	A construct is a mathematical or conceptual model.
Sample	A sample is that part of a population which is actually observed. In normal scientific practice, we demand that it is selected in such a way as to avoid presenting a biased view of the population.
Variance	The variance of a random variable is a measure of its statistical dispersion, indicating how far from the expected value its values typically are. The variance of a real-valued random variable is its second central moment, and it also happens to be its second cumulant.
Normal distribution	The normal distribution is an extremely important probability distribution in many fields. It is a family of distributions of the same general form, differing in their location and scale parameters: the mean and standard deviation. The standard normal distribution is the normal distribution with a mean of zero and a standard deviation of one
Probability distribution	Every random variable gives rise to a probability distribution, containing the most important information about the variable. If X is a random variable, the corresponding probability distribution assigns to the interval (a, b) the probability Pr(a ¡Ü X ¡Ü b), i.e. the probability that the variable X will take a value in the interval (a, b).
Sampling distribution	A sampling distribution is the probability distribution, under repeated sampling of the population, of a given statistic (a numerical quantity calculated from the data values in a sample).
Statistical analysis	Statistical analysis refers to the branch of mathematics that deals with the collection, analysis, interpretation and presentation of masses of numerical data.

Go to **Cram101.com** for the Practice Tests for this Chapter.

Statistic	A statistic is a characteristic of a sample drawn from a population.
Sampling	Sampling is that part of statistical practice concerned with the selection of individual observations intended to yield some knowledge about a population of concern, especially for the purposes of statistical inference.
Probability	Probability is the ratio of the number of favorable outcomes to the number of possible outcomes.
Random sample	A sample is a subset chosen from a population for investigation. A random sample is one chosen by a method involving an unpredictable component, in the sense that the selection of any element of the population is independent of the selection of any other element.
Sample variance	The variance of a random variable is a measure of its statistical dispersion, indicating how far from the expected value its values typically are. The variance of a real-valued random variable is its second central moment, and also its second cumulant. When derived from a sample rather than a population, it is a sample variance.
Sample Mean	The arithmetic mean of a set of numbers is the sum of all the members of the set divided by the number of items in the set. If the set is a statistical population, then we speak of the population mean; if of a sampling of a population, it is a sample mean.
Standard error	The standard error of a measurement, value or quantity is the standard deviation of the process by which it was generated, after adjusting for sample size.
Central limit theorem	The Central Limit Theorem states that if the sum of the variables has a finite variance, then it will be approximately normally distributed. Since many real processes yield distributions with finite variance, this explains the ubiquity of the normal distribution.
Theorem	A theorem is a proposition that has been or is to be proved on the basis of explicit assumptions.
Standard deviation	The standard deviation is the most commonly used measure of statistical dispersion. Simply put, it measures how spread out the values in a data set are. The standard deviation is defined as the square root of the variance.
Deviation	A deviation is the difference between an observed value and the expected value of a variable or function.
Uniform Distribution	In mathematics, the uniform distributions are simple probability distributions. There are two types: the discrete uniform distribution; the continuous uniform distribution.
Standard normal distribution	The standard normal distribution is the normal distribution with a mean of zero and a standard deviation of one (the green curves in the plots to the right). It is often called the bell curve because the graph of its probability density resembles a bell.
Covariance	Intuitively, covariance is the measure of how much two variables vary together. That is to say, the covariance becomes more positive for each pair of values which differ from their mean in the same direction, and becomes more negative with each pair of values which differ from their mean in opposite directions.
Skewness	Skewness is a measure of the asymmetry of the distribution of a real-valued random variable. Skewness, the third standardized moment, is written as $\gamma 1$ and defined as $\gamma 1 = \mu^3 / \sigma^3$ where μ^3 is the third moment about the mean and σ is the standard deviation.
Expected number	Expected number is a term used in probability theory to denote the likely number of trials to occur before a designated event happens. It is the average number that will be derived when the experiment is repeated a large number of times.
Replication	Replication is repeating the creation of a phenomenon, so that the variability associated with the phenomenon can be estimated.

Go to **Cram101.com** for the Practice Tests for this Chapter.

Null hypothesis	A null hypothesis, H_0, is a hypothesis set up to be nullified or refuted in order to support an alternative hypothesis. When used, the null hypothesis is presumed true until statistical evidence in the form of a hypothesis test indicates otherwise.
Hypothesis	A hypothesis is a proposed explanation for a phenomenon. A scientific hypothesis must be testable and based on previous observations or extensions of scientific theories.
Critical value	A critical value is the value corresponding to a given significance level. This cutoff value determines the boundary between those samples resulting in a test statistic that leads to rejecting the null hypothesis and those lead to a decision not to reject the null hypothesis.
Interaction	Interaction is a kind of action which occurs as two or more objects have an effect upon one another. The idea of a two-way effect is essential in the concept of interaction instead of a one-way causal effect.
Proportional	Two quantities are called proportional if they vary in such a way that one of the quantities is a constant multiple of the other, or equivalently if they have a constant ratio.
Residual	Error is a misnomer; an error is the amount by which an observation differs from its expected value; the latter being based on the whole population from which the statistical unit was chosen randomly. A residual, on the other hand, is an observable estimate of the unobservable error.
Sum of squares	Sum of squares is a concept that permeates much of inferential statistics and descriptive statistics. More properly, it is "the sum of the squared deviations". Mathematically, it is an unscaled, or unadjusted measure of dispersion. When scaled for the number of degrees of freedom, it becomes the variance, the sum of squares per degree of freedom.
Histogram	A histogram is a graphical display of tabulated frequencies. That is, a histogram is the graphical version of a table which shows what proportion of cases fall into each of several or many specified categories. The categories are usually specified as nonoverlapping intervals of some variable. The categories (bars) must be adjacent.
Degrees of freedom	In fitting statistical models to data, the vectors of residuals are often constrained to lie in a space of smaller dimension than the number of components in the vector. That smaller dimension is the number of degrees of freedom for error.
Discrete random variable	A probability distribution is called discrete, if it is fully characterized by a probability mass function. Thus, the distribution of a random variable X is discrete, and X is then called a discrete random variable, if: $\sum Pr(X = \mu) = 1$, as u runs through the set of all possible values of X.
Binomial	In elementary algebra, a binomial is a polynomial with two terms: the sum of two monomials. It is the simplest kind of polynomial.
Bernoulli trial	In the theory of probability and statistics, a Bernoulli trial is an experiment whose outcome is random and can be either of two possible outcomes, called "success" and "failure."
Range	In descriptive statistics, the range is the length of the smallest interval which contains all the data. It is calculated by subtracting the smallest observations from the greatest and provides an indication of statistical dispersion.
Correlation	Correlation indicates the strength and direction of a linear relationship between two random variables. In general statistical usage, correlation refers to the departure of two variables from independence.
Multiple regression	A multiple regression is a linear regression with more than one covariate (predictor variable). It can be viewed as a simple case of canonical correlation. An equation used to predict a dependent variable, y from two independents, u and v is: $y = \beta_0 + \beta_1 u + \beta_2 v + \beta_3 u2 + \beta_4 uv + \beta_5 v2$

69

Power	The power of a statistical test is the probability that the test will reject a false null hypothesis, or in other words that it will not make a Type II error. The higher the power, the greater the chance of obtaining a statistically significant result when the null hypothesis is false.
Conditionally independent	In probability theory, two events A and B are conditionally independent given a third event C precisely if the occurrence or non-occurrence of A and B are independent events in their conditional probability distribution given C.
Prediction interval	In statistics, a prediction interval bears the same relationship to a future observation that a confidence interval bears to an unobservable population parameter.
Measurement	Measurement generally refers to the process of estimating or determining the ratio of a magnitude of a quantitative property or relation to a unit of the same type of quantitative property or relation.
Linear regression	In statistics, linear regression is a method of estimating the conditional expected value of one variable y given the values of some other variable or variables x. The variable of interest, y, is conventionally called the "response variable".
Constants	In mathematics and the mathematical sciences, constants are fixed, but possibly unspecified, values.
Confidence Interval	A confidence interval is an interval between two numbers, where there is a certain specified level of confidence that a population parameter lies.
Interquartile range	In descriptive statistics, the interquartile range (IQR) is the difference between the third and first quartiles and is a measure of statistical dispersion. The interquartile range is a more stable statistic than the range, and is often preferred to that statistic.
Quartile	In descriptive statistics, a quartile is any of the three values which divide the sorted data set into four equal parts, so that each part represents 1/4th of the sample or population.
Percentile	In descriptive statistics, a percentile is any of the 99 values that divide the sorted data into 100 equal parts, so that each part represents 1/100th (or 1%) of the sample or population.
Coefficient of Variation	In probability theory and statistics, the coefficient of variation (C_v) is a measure of dispersion of a probability distribution. It is defined as the ratio of the standard deviation to the mean : $C_v = \sigma / \mu$
Bias	A bias is a prejudice in a general or specific sense, usually in the sense for having a predilection to one particular point of view or ideological perspective. However, one is generally only said to be biased if one's powers of judgement are influenced by the biases one holds, to the extent that one's views could not be taken as being neutral or objective, but instead as subjective.
Alternative hypothesis	The alternate hypothesis, or alternative hypothesis, together with the null hypothesis are the two rival hypothesis whose likelihoods are compared by a statistical hypothesis test. Usually the alternate hypothesis is the possibility that an observed effect is genuine and the null hypothesis is the rival possibility that it has resulted from random chance.
Type I error	A false positive, also called a Type I error, exists when a test incorrectly reports that it has found a result where none really exists.
Sample with replacement	Sample with replacement means that in each successive trial of an experiment or process, the total number of possible outcomes, and the mix of possible outcomes is not changed by sampling.
Nominal	In nominal measurement numerals are assigned to objects as labels or names. If two entities

Go to **Cram101.com** for the Practice Tests for this Chapter.

have the same number associated with them, they belong to the same category, and that is the only significance that they have. The only comparisons that can be made between variable values are equality and inequality.

Test Statistic

A test statistic is a summary value (often a summary statistic) of a data set that is compared with a statistical distribution to determine whether the data set differs from that expected under a null hypothesis.

Combinations

Combinations are un-ordered collections of unique elements. The order of the elements is not important

Binomial distribution

The binomial distribution is the discrete probability distribution of the number of successes in a sequence of n independent yes/no experiments, each of which yields success with probability p.